POWER UP WITH ENERGY!

ENERGY FROM THE WIND

by Karen Latchana Kenney

Consultant: Beth Gambro
Reading Specialist, Yorkville, Illinois

BEARPORT
PUBLISHING

Minneapolis, Minnesota

Teaching Tips

Before Reading

- Look at the cover of the book. Discuss the picture and the title.

- Ask readers to brainstorm a list of what they already know about the wind. What can they expect to see in this book?

- Go on a picture walk, looking through the pictures to discuss vocabulary and make predictions about the text.

During Reading

- Read for purpose. Encourage readers to think about the wind and energy and the roles they play in our daily lives as they are reading.

- Ask readers to look for the details of the book. What are they learning about wind?

- If readers encounter an unknown word, ask them to look at the sounds in the word. Then, ask them to look at the rest of the page. Are there any clues to help them understand?

After Reading

- Encourage readers to pick a buddy and reread the book together.

- Ask readers to name one reason to use wind for energy and one reason to not use wind. Go back and find the pages that tell about these things.

- Ask readers to write or draw something they learned about energy from wind.

Credits:
Cover and title page, © WDG Photo/Shutterstock; 3, © ErsoyBasciftci/iStock; 5, © Monkey Business Images/Shutterstock; 6–7, © XiXinXing/iStock; 8–9, © The Vine Studios/Shutterstock; 11, © Spiderstock/iStock; 13, © Space-kraft/Shutterstock; 14, © in4mal/iStock; 16–17, © Volodymyr Burdiak/Shutterstock; 19, © tsmarkley/iStock; 20–21, © franswillemblok/iStock;22, © Naeblys/iStock, © xxmmxx/iStock; 23BL, © Wildroze/iStock; 23BR, © Ron and Patty Thomas/iStock; 23TL, © mbudley/iStock; 23TR, © AlexandrBognat/iStock

Library of Congress Cataloging-in-Publication Data

Names: Kenney, Karen Latchana, author.
Title: Energy from the wind / Karen Latchana Kenney.
Description: Minneapolis, Minnesota : Bearport Publishing Company, [2022] |
Series: Power up with energy! | Includes bibliographical references and
index.
Identifiers: LCCN 2021001064 (print) | LCCN 2021001065 (ebook) | ISBN
9781647478698 (library binding) | ISBN 9781647478766 (paperback) | ISBN
9781647478834 (ebook)
Subjects: LCSH: Wind power--Juvenile literature.
Classification: LCC TJ163.2 .K455 2022 (print) | LCC TJ163.2 (ebook) |
DDC 621.31/2136--dc23
LC record available at https://lccn.loc.gov/2021001064
LC ebook record available at https://lccn.loc.gov/2021001065

For more information, write to Bearport Publishing, 5357 Penn Avenue South, Minneapolis, MN 55419.
Printed in the United States of America.

Contents

Movie Time

Let's watch a movie.

Get cozy in front of the TV.

But what makes the TV turn on?

It may be the wind!

Energy makes things work.

It gives them power.

Our TV turns on because of energy.

And we can get energy from wind.

Wind is moving air.

It pushes things along.

A kite flies in the sky with the wind's energy.

How do we get the wind's energy to use?

It starts with a wind **turbine**.

The turbine is tall and has long **blades**.

Wind spins the blades.

11

The blades move a machine that makes power.

Then, we can use it.

We can make more power with more turbines.

A **wind farm** has lots of them.

A wind farm

We use the energy from wind in many ways.

It powers homes and offices.

It lets us turn on TVs and lights.

Wind blows all around Earth.

It will never run out.

And wind is free.

But it can be hard to get the wind's power.

Some days the wind is not blowing.

Other days are too windy to use turbines.

Still, we are using more wind energy every year.

One day, more power will come from wind.

Let's power up!

Energy from Wind

Follow along as wind becomes power.

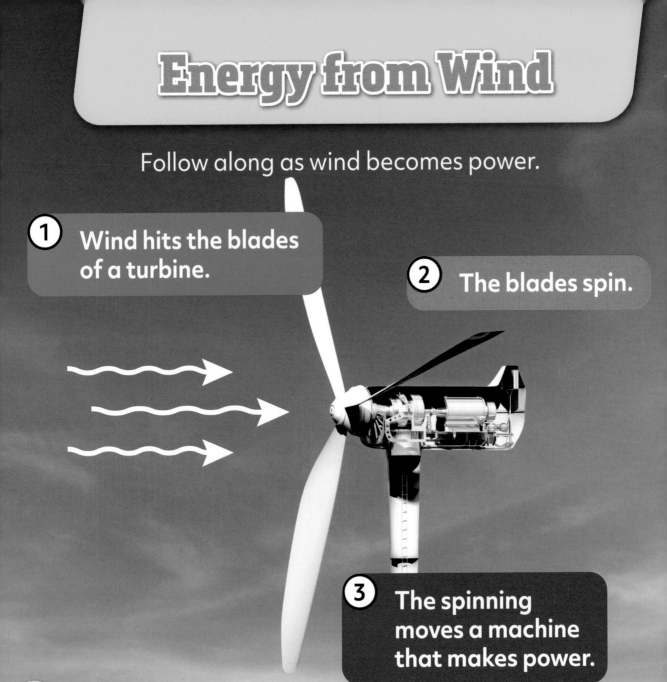

1 Wind hits the blades of a turbine.

2 The blades spin.

3 The spinning moves a machine that makes power.

Glossary

blades long, flat pieces on a turbine that move in the wind

energy power that makes things work

turbine a machine that uses wind to make power

wind farm an area of land with a group of wind turbines

Index

air 8
blades 10–12, 22
kite 8
machine 12, 22

turbine 10–12, 18, 22
TV 4, 6, 15
wind farm 12–13

Read More

Alkire, Jessie. *Wind Energy Projects: Easy Energy Activities for Future Engineers! (Earth's Energy Experiments)*. Minneapolis: Abdo Publishing, 2019.

Felix, Rebecca. *Wind Energy (Earth's Energy Resources)*. Minneapolis: Abdo Publishing, 2019.

Learn More Online

1. Go to **www.factsurfer.com**
2. Enter "**Wind Energy**" into the search box.
3. Click on the cover of this book to see a list of websites.

About the Author

Karen Latchana Kenney likes biking and reading. She tries to find ways to use less energy every day.